AVIAN FLU

2024

Steve k. Bryant

Copyright

Table of Contents

Introduction

An persistent and serious threat to world health is the emergence of infectious diseases. Their erratic behavior and quick dissemination have the potential to have detrimental effects on society and the economy in addition to health. To minimize the impact of these diseases on human populations and to develop effective preventative and control techniques, it is imperative to have a thorough understanding of these diseases. The highly pathogenic H5N1 variant of influenza virus highlights the critical necessity for thorough study and readiness when dealing with such threats. The goal of this book is to give a thorough examination of the H5N1 virus, including its origins, present state, and countermeasures.

Comprehending New Infectious Diseases Is Crucial

Infections that have just surfaced in a population or that have long been present but are expanding quickly in frequency or geographic scope are referred to as emerging infectious diseases. It is crucial to comprehend these illnesses since they pose particular difficulties and may have disastrous consequences.

First, there is a considerable risk of mortality and morbidity from newly developing infectious diseases. The massive COVID-19 pandemic, the Zika virus outbreak in the Americas, and the Ebola outbreaks in West Africa serve as vivid reminders of just how swiftly and widely infectious diseases may spread, overwhelming health systems and igniting widespread panic and disruption.

Second, the economic toll that chronic illnesses frequently take is substantial. Breakouts can cause significant financial losses by interfering with trade, travel, and other economic operations. For example, the 2003 SARS pandemic caused an estimated $40 billion in costs to the world economy. The COVID-19 epidemic has had an even more significant negative impact on the world economy, affecting almost every industry.

Thirdly, creating successful public health interventions requires an awareness of newly emerging infectious illnesses. Controlling outbreaks requires early identification, quick action, and efficient containment techniques. This necessitates a thorough comprehension of the clinical characteristics, social and environmental elements that promote the disease's development, and the dynamics of disease transmission.

And last, new infectious diseases may have an effect on the security of world health. They draw attention to how interrelated the globe is and how important it is for nations to stand together and cooperate. Coordination of efforts between nations, international organizations, and different industries, such as health, agriculture, and wildlife, is necessary for the effective management of these illnesses.

One example of a newly emerging infectious disease that requires our attention is the H5N1 bird flu. It is a major public health concern due to its ability to inflict severe sickness and death, damage agricultural and food security, and mutate into a pandemic strain.

An Overview of the Threat of H5N1 Avian Flu.

A particular strain of the influenza A virus is the cause of the H5N1 avian flu, commonly referred to as bird flu. It is mainly harmful to birds, particularly chickens, though it can also infect people and other animals. In 1996, the virus was discovered in geese in China's Guangdong Province. Since then, it has dispersed over the world, resulting in multiple epidemics in wild and poultry birds.

Because H5N1 has a high pathogenicity and the potential to cause serious illness in people, it is very concerning. Birds infected with the virus frequently pass away in a matter of days, while human mortality rates can reach 60%. Although direct or indirect contact with infected birds has been connected to the majority of human cases, the virus has not yet demonstrated an easy way to transfer from person to person. Since influenza viruses are known to change and reassort, there is a possibility that H5N1 will eventually have the ability to transmit itself continuously from person to person and possibly cause a pandemic.

Agriculture has been significantly impacted by the virus, especially in nations with major chicken industries. Millions of birds may need to be culled in response to outbreaks in poultry in order to stop the virus from spreading, which would cause significant financial losses. Furthermore, consumer behavior may be impacted by the virus's fear, which could lower demand for poultry goods and worsen the economic effects.

Infected and at-risk birds have been culled, farm biosecurity has been strengthened, and surveillance has been carried out to identify and quickly contain outbreaks. In some places, poultry vaccination has also been employed, however there are drawbacks to this strategy, such as the possibility that vaccinated birds could still harbor and spread the virus.

Overview of Avian Influenza

The term "bird flu," or "avian influenza," describes influenza viruses that have been modified for usage in birds. Although avian influenza viruses come in a variety of subtypes, the highly pathogenic avian influenza (HPAI) viruses, including H5N1, are the most important to human health.

Hemagglutinin (H) and neuraminidase (N), two proteins on the surface of influenza viruses, are used to classify them. These proteins are essential to the virus's capacity to propagate and infect host cells. There are 11 neuraminidase subtypes (N1 to N11) and 18 recognized hemagglutinin subtypes (H1 to

H18). Worldwide, wild birds are known to carry avian influenza viruses, with water species such as ducks and geese serving as natural hosts for the viruses.

Some avian influenza viruses, such as H5N1, can cause severe sickness and significant mortality rates, especially in domestic poultry. Most avian influenza viruses cause mild disease in birds. These extremely dangerous diseases can spread quickly across flocks, posing serious risks to food security and the economy.

Direct contact with infected birds or polluted settings can result in human infection with avian influenza viruses. While infrequent, human infections can be quite serious, especially when it comes to H5N1. Human symptoms can vary from the common ones associated with the flu (fever, cough, sore throat, muscle aches) to more serious respiratory conditions like pneumonia and acute respiratory distress syndrome (ARDS), which can be life-threatening.

Because avian influenza viruses can mutate and reassort, there is a chance that they will create pandemics. Because influenza viruses have segmented genomes, if a host is co-infected with more than one strain of the virus, the viruses can swap genetic material. This may cause new viruses to evolve that have never been seen before, including the capacity to infect humans and transmit quickly from one person to another.

The Historical Background of H5N1 Pandemics

The H5N1 avian flu has a history of recurrent outbreaks and persistent worries about the virus's ability to spread over the world. In 1996, the virus was discovered in geese in China's Guangdong Province. It led to an outbreak in Hong Kong's poultry the following year, marking the first instance in which H5N1 was linked to a serious illness in both humans and animals. Six of the eighteen human instances that were recorded resulted in death. Poultry was mass-culled in order to contain this outbreak.

H5N1 persisted in causing occasional outbreaks across Asia in the ensuing years. But in 2003–2004, the virus started to spread more broadly, leading to serious outbreaks in poultry in several Asian nations, including Vietnam, Indonesia, and Thailand. It was during this time that H5N1 started to gain international attention.

H5N1 was discovered in migratory birds in 2005 near Qinghai Lake in China, indicating the virus's potential for long-distance transmission. Soon after, epidemics in birds were documented in Africa, the Middle East, and Europe. Human cases of the virus persisted; these cases mostly affected those who had close contact with sick birds.

Both national and international efforts have been made in response to H5N1. In order to quickly discover and contain epidemics, countries that have experienced them have put in place measures like culling sick and at-risk birds, improving biosecurity on farms, and performing surveillance. These initiatives have received direction and assistance from international agencies such as the World Health Organization (WHO), the Food and Agriculture Organization (FAO), and the World Organization for Animal Health (OIE).

The H5N1 virus continues to pose a hazard despite these precautions. The virus is still spreading among wild birds and poultry, and there are still occasional reports of human cases. The necessity for constant surveillance is highlighted by the high human mortality rate and the possibility that the virus will mutate into a strain that can transmit itself effectively from person to person.

The possibility of a pandemic caused by H5N1 is one of the biggest worries. When a novel influenza virus that can infect people and spread quickly from person to person appears, it can cause influenza pandemics. The 1918 Spanish flu pandemic, which is thought to have killed 50 million people globally, was the most catastrophic in recent memory. The 2009 H1N1 pandemic, like other more recent pandemics, brought to light the influenza viruses' quick spread and widespread effects.

Despite the fact that H5N1 has not yet demonstrated the ability to spread quickly among humans, its high death rate and potential for genetic modification make it a major threat. The potential for the virus to acquire mutations that facilitate effective transmission from person to person is still a major area of study and concern for preparedness.

To sum up, the H5N1 avian flu poses a serious and persistent risk to international health. Creating successful plans to stop and manage the virus's transmission requires a thorough understanding of the infection, its history, and its possible effects. This book seeks to give readers a thorough understanding of the H5N1 virus, covering everything from its virology and epidemiology to the public health precautions that are required to lessen its effects. We aim to contribute to the ongoing efforts to safeguard public health and avert future pandemics by looking at previous outbreaks, ongoing research, and upcoming preparedness initiatives.

Chapter 1

The H5N1 Science

The H5N1 virus, a subtype of the influenza A virus, has attracted a lot of attention since it can cause serious sickness in people and has a high pathogenicity in birds. Creating efficient preventative and control plans requires a thorough understanding of the virology, genetic composition, modes of transmission, and host range of H5N1. Furthermore, contrasting H5N1 with other influenza viruses draws attention to its distinct features and the particular difficulties it presents.

H5N1's Virology and Genetic Composition

The influenza A, B, and C viruses are members of the Orthomyxoviridae family, which also includes the H5N1 virus. Hemagglutinin (H) and neuraminidase (N), two surface glycoproteins, are combined to further classify influenza A viruses. There are 11 neuraminidase subtypes (N1 to N11) and 18 recognized hemagglutinin subtypes (H1 to H18). Thus, the N1 neuraminidase and H5 hem agglutinin are present in the H5N1 virus.

The flu The genome of a virus is divided into eight segments of single-stranded RNA. Because of their segmented structure, influenza viruses can swap genetic material with one another when they infect the same host cell, a process known as genetic reassortment. New viral strains with unique characteristics, such as variations in virulence, transmissibility, and host range, may arise from this reassortment.

The protein known as hemagglutinin (H) is essential for the virus to infect host cells. It facilitates viral entrance by binding to sialic acid receptors on the surface of host cells. Birds' respiratory and digestive tracts are home to the majority of avian-type receptors, which the H5 hemagglutinin subtype is especially affinious for. However, changes in the specificity of receptor binding caused by mutations in the hemagglutinin protein may make it possible for the virus to infect humans and other species.

When new viral particles are released from infected cells, the neuraminidase (N) protein is essential. It breaks down sialic acid residues to stop viral particles from clumping together and from spreading to other cells. The capacity of the H5N1 virus to produce severe illness and high rates of mortality in poultry has been linked to its high pathogenicity in birds, specifically in relation to its N1 neuraminidase.

There is a genetic marker association with the high pathogenicity of H5N1 in birds. The presence of several basic amino acids at the hemagglutinin protein's cleavage site is one such indication. The hemagglutinin protein can be broken by a wider variety of host proteases thanks to this polybasic

cleavage site, which makes it easier for the virus to travel from the gastrointestinal and respiratory tracts to other organs and cause systemic infection.
Phylogenetic research of the hemagglutinin gene has revealed other clades and subclades within the H5N1 virus, indicating further genetic diversity in the virus. The pathogenicity, transmissibility, and antigenic features of the virus can all be affected by these genetic variants, which makes it more difficult to create vaccines and control strategies that work.

Modes of Transmission and Range of the Host

The H5N1 virus primarily infects wild aquatic birds, including swans, ducks, and geese. These birds serve as reservoirs, carry the virus frequently asymptomatically, and are essential to the ecology and epidemiology of the virus. Direct contact, contaminated water sources, and fecal-oral pathways are the three ways that the virus can spread among wild birds.

Chickens, turkeys, and ducks are among the domestic poultry that are most vulnerable to H5N1 infection. When sick birds come into close contact with one another or with polluted feed, water, or equipment, the virus can spread quickly among flocks. Poultry outbreaks have the potential to cause significant mortality rates, making the culling of entire flocks necessary to stop the virus's transmission.

Although they are uncommon, direct or indirect contact with diseased birds or polluted settings can result in H5N1 infections in humans. There are a lot of risks involved in tasks including handling sick or dead birds, killing, defeathering, and preparing chicken for eating. Severe respiratory sickness can result from the virus entering the human body through the conjunctiva, respiratory system, or mucous membranes.

H5N1 does not just infect people and birds as hosts. Numerous animal species, including dogs, cats, tigers, and pigs, have been found to harbor the virus. Because they can co-infect with avian, human, and swine influenza viruses and act as "mixing vessels" for reassortment events that could result in the production of novel strains with pandemic potential, pigs are especially dangerous.

Sustained human-to-human transmission of H5N1 has not been reported, despite its wide host range. The majority of human cases have been connected to intimate contact with infected birds, and a small number of cases—usually involving close family members or medical personnel—have shown evidence of limited human-to-human transmission. The virus prefers avian-type receptors, which are less common in humans, which contributes to its inefficiency in spreading among them.

H5N1 Variations From Other Influenza Viruses

A comparison of H5N1 with other influenza viruses reveals a number of significant distinctions concerning pathogenicity, transmissibility, and public health impact. These viruses include seasonal influenza A viruses (H1N1 and H3N2) as well as the 2009 H1N1 pandemic virus.

Pathogenicity

H5N1 is a highly pathogenic virus that frequently causes severe illness and a high death rate in birds. On the other hand, seasonal influenza A viruses usually cause mild to severe sickness with a lower death rate in humans and birds.

Seasonal influenza viruses typically have mortality rates of less than 0.1%, while H5N1 infections in humans are linked to a high case-fatality rate of about 60%.

Genetic indicators, such as the polybasic cleavage site in the hemagglutinin protein, which is missing in the majority of seasonal influenza viruses, are connected to the high pathogenicity of H5N1.

Transferability

Humans are very susceptible to seasonal influenza viruses, which mostly spread through respiratory droplets and, to a lesser extent, by fomites and aerosols. These viruses have evolved to connect effectively to upper respiratory tract human-type receptors, which makes person-to-person transmission easier.

However, H5N1 typically binds to receptors of the avian type, which are less common in the upper respiratory tract of humans. Because of its receptor specificity, it cannot propagate effectively among humans.

A few incidences of restricted human-to-human H5N1 transmission have been reported, but they are uncommon and have not resulted in long-term community transmission.

Genetic Variability and Mixing

Influenza A virus's segmented genome, which permits reassortment, allows for a high degree of genetic variety. Novel features of new virus strains may arise as a result of this process.

The seasonal flu vaccination needs to be updated on a regular basis because seasonal influenza viruses frequently experience antigenic drift, which leads to minute genetic alterations.

Although H5N1 is also susceptible to genetic drift and reassortment, its high pathogenicity and ability to acquire mutations that could increase its transmissibility among people make it a particularly concerning virus. Attempts to create efficient vaccinations and containment strategies are hampered by the formation of new H5N1 clades and subclades.

Impact on Public Health

Seasonal influenza viruses create yearly epidemics that inflict considerable morbidity and mortality on a global scale, especially in susceptible groups like the elderly, small children, and people with underlying medical disorders.

The 2009 H1N1 pandemic virus was a worldwide health and economic disaster that resulted from a reassortment of avian, swine, and human influenza viruses. The virus's effective human-to-human transmission allowed it to spread quickly, although compared to H5N1, its death rates were lower.

Although H5N1 hasn't started a pandemic, there is still reason for concern that it might. The high rate of human case fatalities along with the potential for genetic alterations to increase transmissibility highlight the necessity of continued research, preparation, and surveillance.

Control Strategies

Seasonal influenza can be controlled with annual vaccinations, antiviral drugs, and public health campaigns that encourage good hand hygiene and respiratory manners.

Control strategies for H5N1 in poultry are essential to avoiding infections in humans. These precautions include removing diseased and vulnerable birds from the flock, improving farm biosecurity, and monitoring to quickly identify and contain epidemics.

In certain regions, vaccination of poultry has been employed to manage H5N1, but there are still issues, such as the possibility that vaccinated birds could harbor and transmit the virus.

To sum up, the H5N1 virus poses distinct difficulties in contrast to other influenza viruses. It is a major public health concern due to its high pathogenicity in birds, potential for serious disease in people, and ability to mutate genetically. For the purpose of creating efficient plans to stop and manage the spread of H5N1, it is imperative to comprehend the virus's virology, genetic composition, modes of transmission, and host range. Furthermore, contrasting H5N1 with other influenza viruses brings to light the unique traits and dangers connected to this virus, emphasizing the necessity of continued watchfulness and readiness to lessen the effects of upcoming epidemics.

Chapter 2

The Chronicle of H5N1 Pandemics

Early Speaks and First Reactions

In 1996, the H5N1 avian flu virus was discovered in China's Guangdong Province in geese. After it was first discovered, a number of outbreaks that would eventually infect both humans and poultry worldwide began. The virus's capacity to infect humans and cause serious illness in birds raised serious concerns, which drove the adoption of a number of control measures.

Hong Kong Outbreak in 1997

Hong Kong saw one of the first and most significant H5N1 outbreaks in 1997. This was the first time that H5N1 had been seen to cause a serious illness in humans. The outbreak started in chickens and quickly spread to humans despite species barriers. Six fatalities were reported from a total of 18 human cases. Public health professionals were concerned by the high case-fatality rate and took immediate action to limit the infection.

Over the course of three days, some 1.5 million chickens and other fowl were mass-culled as part of Hong Kong's reaction. This bold approach effectively contained the pandemic and stopped new human cases. But the epidemic in Hong Kong in 1997 showed that H5N1 might result in serious illness and even death in humans, underscoring the importance of being alert and ready.

Asian Outbreaks in the Early 2000s

Following the 1997 outbreak, H5N1 persisted in spreading among poultry in some Asian regions, occasionally leading to human illnesses. But the virus didn't start to spread more broadly until the early 2000s, at which point it caused a number of outbreaks in various Asian nations.

H5N1 caused outbreaks in Japan and South Korea's poultry populations in 2003. The virus made its way to Vietnam, Thailand, Indonesia, and Cambodia the next year. High rates of mortality in poultry and sporadic human cases—usually involving people who had close contact with diseased birds—were the defining characteristics of these outbreaks.

Countries responded differently to these outbreaks, but in general, farms improved biosecurity measures, culled diseased and at-risk chickens, and increased surveillance to quickly identify and contain epidemics. The virus persisted in spreading in spite of these attempts, underscoring the difficulties in containing H5N1 in areas with substantial numbers of poultry and few resources.

Significant Pandemics in History and Their Effects

Examining past influenza pandemics is necessary to comprehend the possible effects of H5N1. Although H5N1 hasn't started a pandemic yet, its traits make it possible that it could eventually develop the capacity to spread quickly among people. Important lessons about the possible outcomes of such an event can be learned from historical pandemics.

Spanish Flu of 1918

Most people agree that the 1918 Spanish flu pandemic was the worst influenza outbreak in recorded history. The Spanish flu, which killed an estimated 50 million people worldwide, affected one-third of the world's population and was caused by the H1N1 influenza A virus. The high death rate of the pandemic—particularly among young adults—and its quick spread brought to light the serious effects an influenza virus can have on society and world health.

The Spanish flu caused severe disruptions to daily life, overtaxed healthcare institutions, and resulted in large financial losses, among other social and economic effects. The pandemic also made clear how crucial it is to implement public health precautions including face mask use, isolation, and quarantine in order to stop the spread of dangerous diseases.

The Hong Kong flu in 1968 and the Asian flu in 1957

An H2N2 influenza A virus was the cause of the 1957 Asian flu pandemic. It started in East Asia and quickly spread over the world, killing between one and two million people. Genetic reassortment between an avian influenza virus and a human influenza virus led to the creation of the H2N2 virus, underscoring the possibility that reassortment between avian influenza viruses could cause pandemics.

The creation and dissemination of a vaccine as part of the reaction to the Asian flu served to lessen the pandemic's effects. The Asian flu experience highlighted the value of international collaboration, vaccine development, and surveillance in the fight against influenza pandemics.

Similar patterns were seen in the 1968 Hong Kong flu pandemic, which was brought on by the H3N2 influenza A virus. A human influenza virus and an avian influenza virus reassorted to create the H3N2 virus. Globally, the pandemic is thought to have killed one to four million people. The creation of a vaccine was part of the reaction to the Hong Kong flu and was essential in halting the virus's spread.

The H1N1 Pandemic of 2009

2009 saw the most recent influenza pandemic, which was brought on by a novel H1N1 influenza strain. a virus with genetic components from swine, human, and avian influenza viruses that first appeared in pigs. The swine flu pandemic of 2009, also called the H1N1 pandemic, spread quickly and infected 24-28% of the world's population in its first year.

Despite its high transmission rate, the 2009 H1N1 virus mostly caused mild to severe disease and was responsible for an estimated 200,000 fatalities globally. The quick creation and release of a vaccine, the use of antiviral drugs, and public health initiatives to stop the spread of the virus were all part of the reaction to the pandemic.

The 2009 H1N1 pandemic brought to light the significance of being ready for pandemics, which includes having efficient surveillance systems, the ability to produce vaccines quickly, and the ability to coordinate public health responses. It also highlighted the fact that influenza viruses, even those that are not very dangerous, have the capacity to have a major influence on world health.

The H5N1 Strain's Evolution Over Time

Significant genomic change has occurred in the H5N1 virus since its discovery in 1996. The virus's segmented genome, which permits genetic reassortment, and the mutations that take place during viral replication have been the main drivers of its evolution. As a result, there are numerous

H5N1 clades and subclades, each with distinct genetic and antigenic characteristics.

Early Genetic Alterations and Their Dispersal

Following the discovery of the first H5N1 virus in 1996, multiple genetically different lineages emerged. As a result of the virus's spread from China to other Asian countries, several clades were established. The hemagglutinin gene, a crucial factor in determining the virus's antigenic characteristics and host range, varied among these clades.

The acquisition of a polybasic cleavage site in the hemagglutinin protein was one of the most important genetic modifications throughout the early evolution of H5N1. This mutation enhanced the pathogenicity of the virus in birds by enabling a wider variety of host proteases to break the hemagglutinin, so enhancing systemic infection.

The Clade 2.3.4.4 Emerges

Clade 2.3.4.4 is one of the most prominent H5N1 clades; it first appeared in the middle of the 2000s.

This clade has been linked to sporadic human infections as well as extensive epidemics in wild birds and poultry. Multiple subclades of the virus known as Clade 2.3.4.4 have been found, indicating a significant degree of genetic variation in the group. The development of vaccines is hampered by the notable antigenic variation shown by Clade 2.3.4.4. The hemagglutinin protein's antigenic characteristics are subject to change over time, which can diminish the efficacy of current vaccinations and require the creation of new vaccine candidates.

Expanded throughout the Middle East, Africa, and Europe

H5N1 was found in migrating birds at China's Qinghai Lake in 2005. This was the first sign that the virus was spreading outside of Asia. A major factor in the spread of H5N1 throughout Europe, Africa, and the Middle East was migratory birds. The virus was found in wild birds and poultry in a number of nations, resulting in large epidemics and financial damages.

The difficulties in managing the virus in places with different agricultural practices and different degrees of public health infrastructure were brought to light by the H5N1 outbreak's expansion into new locations. It also emphasized how crucial international coordination and collaboration are when reacting to diseases that affect animals across borders.

Current Situation and Recent Developments

New clades and subclades of H5N1 have emerged in recent years as the virus has continued to evolve. The virus is still widespread in poultry across a number of nations, and outbreaks are still happening on occasion. Although they are uncommon, human cases are still documented; these cases usually involve people who have had close contact with diseased birds.

A primary apprehension regarding the continuous development of H5N1 is the possibility of the virus obtaining alterations that augment its ability to spread among humans. Numerous mutations have been found to boost the binding affinity of the hemagglutinin protein for human-type receptors and to enhance viral replication in mammalian cells. These mutations have been linked to various health issues.

Surveillance systems that track the genetic and antigenic characteristics of the virus in both poultry and wild birds are part of the efforts to monitor the evolution of H5N1. The development of vaccines and other control measures, as well as the identification of newly developing strains with pandemic potential, depend on these initiatives.

In summary many difficulties have been encountered during H5N1 outbreaks, and worries about the virus's ability to start a worldwide pandemic have persisted. The virus's great pathogenicity in birds and its capacity to infect humans were brought to light by early outbreaks in Asia, prompting important public health measures. Significant past pandemics, like the Spanish flu of 1918 and the H1N1 pandemic of 2009, teach valuable lessons about the possible consequences of influenza viruses and the need of being ready.

The H5N1 virus has evolved throughout time into a wide variety of clades and subclades, each with distinct genetic and antigenic characteristics. The migrating birds' assistance in the virus's transmission to new areas.

Chapter 3

Initial Results from the CDC

Summary of the CDC's Current Investigations and Results

The H5N1 avian flu virus has been extensively studied and tracked by the Centers for Disease Control and Prevention (CDC), with particular attention paid to the virus's pathogenicity, immunity levels, and dissemination throughout the US population. The CDC has increased its efforts to comprehend the possible harm posed by this virus in light of recent outbreaks on dairy and chicken farms. In order to ascertain the present state of the virus and its potential effects on public health, their job entails comprehensive testing and analysis of samples from both humans and animals.

The monitoring of the virus in bird populations, the tracking of human infections, and serological investigations to evaluate the immune response in the population are all important components of the CDC's continuous testing program. To maintain a thorough approach to H5N1 surveillance, the CDC works in conjunction with other federal agencies, state health departments, and foreign organizations. Blood sample collection and analysis from different parts of the country has been a major part of the CDC's activities. Antibodies against the H5N1 virus have been tested on these samples in order to determine the degree of pre-existing immunity among Americans. The results of this serological testing are essential for determining the population's possible level of readiness in the case of a pandemic or a more widespread epidemic.

Results of the Serology Study for 2021–2022 and 2022–2023

Using blood samples obtained from people in all 10 U.S. regions during the flu seasons of 2022–2023 and 2021–2022, the CDC carried out serological investigations. The purpose of these investigations was to find the predominance of antibodies against the H5N1 virus, as this would reveal prior exposure or immunity brought on by vaccination.

Techniques

In order to determine if antibodies existing in the blood could neutralize the virus, scientists challenged blood samples with live H5N1 virus as part of a procedure known as virus neutralization assays. This technique offers a trustworthy indicator of whether the immune system has already come into contact with the virus and is capable of mounting a defense.

Results

These serology investigations' findings showed that there were extremely low levels of H5N1 antibodies everywhere. This was true whether or not people had received the seasonal flu vaccine, which shares antigens with several flu strains and may provide some cross-reactive immunity but is not intended to protect against H5N1.

The great majority of Americans have little to no pre-existing immunity to H5N1, as indicated by the low antibody levels. This means that if the virus were to change and become more easily transmissible among humans, the majority of people would be vulnerable to infection.

The American population's low pre-existing immunity and its implications

Low levels of pre-existing immunity to H5N1 in the American population have been found, which has important ramifications for pandemic preparedness and public health. To reduce the possibility of a possible H5N1 breakout, policies must take these consequences into consideration.

Possibility of Quick Dissemination

One of the most alarming consequences is that if H5N1 were to evolve to become more adept at transmitting from person to person, it might spread quickly. At present, H5N1 mainly affects birds and can also occasionally infect humans, usually those who have intimate contact with poultry that is afflicted. On the other hand, influenza viruses are recognized for their propensity to reassort and evolve, possibly gaining the genetic modifications required for persistent human transmission.

Due to the population's low levels of immunity, an outbreak might spread swiftly and extensively, resulting in a high rate of morbidity and mortality. This situation emphasizes how crucial it is to keep a careful eye out for any indications of the virus becoming more transmissible.

Developing and storing vaccines

The urgent requirement for efficient H5N1 vaccinations is highlighted by the absence of pre-existing immunity. Although there are potential vaccinations out there, they are not yet produced in large enough quantities for general usage. In order to guarantee that production can be swiftly increased in the case of an outbreak, the CDC and other health organizations are collaborating with vaccine makers. Federal health officials signed a contract with CSL Seqirus in late May 2024 to complete the production of one of the two potential H5N1 vaccines in bulk. With the help of this deal, 4.8 million medicines will always be on hand and ready to be quickly distributed as necessary. Important to this vaccine's efficacy is the CDC's confirmation that it closely resembles the H5N1 virus that is currently in circulation.

Antiviral Drugs

Antiviral drugs are essential for treating and managing influenza outbreaks in addition to vaccinations. It has been demonstrated that oseltamivir and zanamivir, two neruriminidase inhibitors, are beneficial in lowering mortality linked to bird flu. A crucial part of pandemic preparedness is making sure that these drugs are readily available and in sufficient quantities.

Public Health Interventions and Monitoring

The low levels of immunity further emphasize the necessity of strong public health initiatives to stop the spread of H5N1. Enhanced biosecurity on farms, culling of diseased poultry, and quarantine and isolation of affected persons are some of these approaches. It is imperative to conduct routine monitoring of animal and human populations in order to immediately identify and contain epidemics. The CDC's findings highlight how crucial international collaboration is to the surveillance and management of avian influenza. Due to migratory bird patterns and the worldwide chicken trade, an outbreak in one area can swiftly spread to other parts of the world and pose a concern. Effective surveillance and response depend on cooperative efforts with agencies like the Food and Agriculture Organization (FAO) and the World Health Organization (WHO).

Education and Public Awareness

Another crucial component of readiness is informing the public about the dangers of H5N1 and the precautions they can take. This entails encouraging frequent hand washing, handling and preparing chicken safely, and preventing contact with dead or ill animals. Campaigns for public awareness can guarantee that people are aware of what to do in the event of an outbreak and help lower the risk of illnesses among humans.

The vulnerability is evident from the CDC's preliminary results regarding the serological status of the American population with regard to the H5N1 avian flu. One cannot discount the possibility of a serious epidemic, or perhaps a pandemic, given the low levels of pre-existing immunity. These results highlight the necessity of ongoing watchfulness, readiness, and quick reaction times.

A comprehensive approach to preserve public health must include efforts to produce and stockpile vaccinations, guarantee the availability of antiviral drugs, and implement efficacious public health measures. To lessen the threats presented by H5N1, public education and international cooperation are also crucial.

Remaining knowledgeable and ready is still the best line of protection against the constant threat of newly developing infectious diseases such as H5N1. This is especially true as the CDC continues its research and monitoring activities.

Chapter 4

Effects on Farms that Raised Dairy and Poultry

Incidences in Dairy Herds and Poultry Flocks That Continue

Dairy herds and poultry flocks have been significantly impacted by the H5N1 avian flu virus, especially in the US. The virus, which is highly harmful in birds, has caused multiple outbreaks all over the nation and is causing significant problems for the agriculture industry. These epidemics have had significant negative effects on cattle health in addition to having broad economic and social ramifications.

Current Circumstance

According to the most recent data, the Animal and Plant Health Inspection Service (APHIS) of the U.S. Department of Agriculture (USDA) has confirmed over 100 H5N1 outbreaks in dairy cows, with additional sporadic detections in poultry flocks. Numerous states have reported experiencing these outbreaks, with Colorado and Iowa suffering the most. The fact that the virus has spread despite efforts to limit it emphasizes how difficult it is to control a sickness that is so highly contagious.

Management of Outbreaks

A variety of tactics are used to manage H5N1 outbreaks in livestock, such as killing exposed and sick animals, enforcing stringent biosecurity regulations, and boosting surveillance to quickly identify new cases. These steps are intended to slow the virus's spread and lessen its effects on the agriculture sector.

Culling has substantial negative effects on farmers' finances and psychological well-being even if it is an efficient way to stop the virus from spreading. Numerous animals are destroyed, resulting in significant financial losses and emotional suffering for all parties concerned.

Improving biosecurity protocols is essential for stopping H5N1 from entering and spreading across farms. These precautions include limiting access to farms, cleaning farm machinery and vehicles, and making sure farm laborers adhere to stringent hygiene regulations.

Particular Case Studies of Current Colorado and Iowa Outbreaks

A closer look at case studies of recent outbreaks in Colorado and Iowa can help shed light on the effects of H5N1 on dairy and poultry farms. These examples highlight the difficulties farmers encounter and the solutions they employ to deal with the situation.

Colorado Epidemics

Numerous dairy farms in Colorado have reported H5N1 epidemics that have impacted thousands of cattle. The dairy industry in the state experienced severe problems as a result of the most recent confirmations, which implicated five farms.

Case Study: Colorado's Weld County Farm

One of the noteworthy outbreaks happened in Weld County on a sizable dairy farm. Cattle deaths at the farm, which housed over 5,000 animals, suddenly increased. After H5N1 was detected through laboratory testing, state and federal authorities acted right away.

Reaction and Management Strategies

Culling and Disposal: In order to stop the virus from spreading further, the exposed and afflicted cattle were culled. Strict biosecurity precautions were followed during the culling procedure to reduce the possibility of virus spread.

Disinfection: To get rid of any leftover virus particles, a thorough disinfection of farm machinery, cars, and buildings was carried out.

Quarantine: The farm was put under quarantine, limiting the flow of people and animals into and out of the property.

Surveillance: To quickly identify any new cases, increased surveillance was put into place throughout the neighborhood.

Effect on Agriculture

The farm owner suffered large financial losses as a result of the outbreak as in addition to losing a sizable number of cattle, there were additional expenses related to surveillance and disinfection. In addition, the loss of their cattle and the uncertainty of what lay ahead took a significant emotional toll on the farmer and the farm laborers.

Outbreaks in Iowa

Another state severely hit by H5N1, Iowa, reported several outbreaks in flocks of chickens as well as dairy cows. The most recent confirmations included ranches in Plymouth County and Sioux County, areas well-known for having a significant concentration of cattle farms.

Case Study: Iowa's Plymouth County Poultry Farm

Thousands of birds in Plymouth County became infected as a result of an epidemic on a sizable chicken farm. The virus caused significant interruptions to the farm, which produced both meat and eggs.

Reaction and Management Strategies

Culling: To stop the infection from spreading, all exposed and diseased birds were put to death. Over 100,000 hens had to be destroyed as a result, which had a big effect on the farm's output.

Disinfection: The farm was thoroughly cleaned, with coops, machinery, and transport vehicles all receiving special attention.

Movement Restrictions: Strict guidelines were put in place for the farm, prohibiting the transportation of grain, eggs, and birds to and from the property.

Enhanced Biosecurity: Stricter hygienic regulations and more surveillance of agricultural laborers were implemented as part of heightened biosecurity measures.

Effect on Agriculture

On the chicken farm, the financial effects were disastrous. There were financial losses as a result of the substantial drop in egg and meat output brought on by the loss of so many hens. The expenses of cleaning, culling, and improved biosecurity procedures also put further burden on the farm's finances.

There was also a significant social influence on the farming community. As a result of the outbreak and the farm's response to lower production, jobs were lost. The psychological effects on farm laborers, who had to put a lot of birds down, were also noteworthy.

The farming industry's effects on the economy and society

There are significant social and economic ramifications to the current H5N1 outbreaks in chicken and dairy farms. These effects affect not just the impacted farms but also the rural communities and the larger agricultural industry.

Financial Affect

H5N1 outbreaks have a complex economic impact on the farming sector, affecting both short- and long-term financial stability.

Specific Monetary Losses

Livestock Losses: Farmers suffer immediate financial losses as a result of culling diseased and exposed animals. This includes the death of productive animals, which lowers the amount of money made from the production of milk, meat, and eggs.

Costs of Biosecurity and Disinfection: Putting improved biosecurity and disinfection measures into place comes with a hefty price tag. Farmers, particularly those with low resources, must bear the financial burden of these measures even if they are crucial for curbing the virus's spread.

Movement and Quarantine Restrictions: These measures cause havoc with the supply chain, making it difficult to transfer grain, cattle, and other necessities. Farmers may experience delays and increased expenses as a result.

Undirect Financial Losses

Market Disruptions: H5N1 outbreaks have the potential to cause market disruptions, such as a decline in the demand for dairy and poultry goods. Fears about product safety may prevent consumers from buying these goods, which would result in lower pricing and less money for farmers.

Trade Restrictions: In reaction to H5N1 outbreaks, trade restrictions may be implemented internationally, impacting the export of dairy and poultry goods. For the agriculture industry, this might mean large financial losses, especially for export-dependent nations.

Insurance and Compensation: It can be difficult for farmers to get insurance coverage for damages brought on by H5N1 outbreaks. Government programs may offer compensation, but it might not be enough to fully offset the degree of the losses.

Social Repercussions

H5N1 outbreaks have a significant social impact on farming and rural communities, impacting not just the individual but also the larger community.

Effect on the Mind

Farmer Stress and Anxiety: The financial losses and unpredictability of the future cause farmers impacted by H5N1 epidemics to feel a great deal of stress and anxiety. Livestock losses can be heartbreaking because they frequently symbolize years of arduous labor and commitment.

Farm Workers' Emotional Toll: Farm workers experience emotional distress during the culling of diseased animals. They may also experience psychological effects from having to put many animals to sleep. Stress levels may rise and mental health problems may result from this.

Community Effect

Job Losses: When farms cut staff due to lower output and financial strain, H5N1 outbreaks may result in job losses. This impacts not just the jobless individuals but also their families and the larger community.

Economic reduction: Rural areas that mostly depend on agriculture may see a reduction in their economies as a result of H5N1 outbreaks. The community's overall economic health may be impacted by lower spending at local businesses as a result of farmers' income declines and job losses.

Social Stigma: Afraid of the virus spreading, farmers and farm workers impacted by H5N1 epidemics may experience social stigma. Within the community, this may result in alienation and loneliness.

H5N1 epidemics have a substantial and varied effect on dairy and poultry farms, influencing rural communities' social cohesion and economic stability. The recurring outbreaks in regions like Iowa and Colorado highlight the difficulties farmers have controlling and combating a disease that is so highly contagious. The emotional and financial costs incurred by farmers and agricultural laborers highlight the necessity of extensive support networks and efficient control strategies.

Prioritizing vaccine development, bolstering biosecurity protocols, and offering impacted farmers emotional and financial support are critical as the CDC and other agencies watch for and respond to H5N1 outbreaks. Future plans to lessen the impact of H5N1 and other emerging infectious illnesses on the rural communities and agriculture sector can be influenced by the lessons acquired from past outbreaks.

Chapter 5

Consequences for Human Health

Public Risk Assessment

The H5N1 avian flu virus primarily affects birds, but because it may cause severe illness and high mortality in infected individuals, it poses a serious risk to human health. Evaluating a number of variables, such as the probability of human infections, the severity of the illness, and the possibility of viral mutation and increased human-to-human transmission, is necessary to comprehend the risk to the broader population.

Present Danger Amounts

Because H5N1 does not spread easily from birds to humans or between humans, there is now little risk to the general public. The majority of human cases have happened to people who came into close contact with diseased birds or polluted areas, like live bird marketplaces or poultry farms. But if the virus experiences genetic alterations that increase its transmissibility among people, things might quickly change.

Factors Affecting Danger

Animal-Human Interface: People who deal closely with poultry, such as farmers, veterinarians, and employees of live bird markets, are most at risk of contracting an infection from them. Due to their close proximity to the infection, these populations are more likely to come into contact with contaminated objects or diseased birds.

Virus Mutations: Influenza viruses are prone to mutation and reassortment, and the H5N1 virus is no exception. Modest genetic alterations can drastically modify the virus's capacity to infect humans and move from one individual to another. Keeping an eye on these genetic alterations is essential to determining the likelihood of a future pandemic.

Public Health Preparedness: One of the most important factors in reducing the risk is the degree of readiness and responsiveness of public health systems. Essential elements of readiness include efficient monitoring, quick diagnostics, antiviral drug availability, and immunization plans.

Public Awareness and Behavior: Spreading knowledge about the dangers of H5N1 and encouraging actions that lessen the risk of contracting the virus, like using appropriate hand

Handling and cleaning chickens properly can greatly reduce the chance of illness in humans. In order to control outbreaks, public adherence to health advisories and preventive actions is crucial.

Human Infection Cases and Their Backgrounds

Despite the comparatively low incidence of H5N1 infections in people, the cases that have happened offer important insights into the circumstances and environments that promote bird-to-human transmission. By analyzing these cases, public health measures aimed at lowering the risk of infection in the future can be informed by the identification of high-risk locations and activities.

Prominent Incidents and Epidemics

Asia: The H5N1 Human Infection Epicenter

Asia has been the region where most human H5N1 infections have happened, especially in nations with high poultry numbers and close human-animal contact. Many cases have been recorded from China, Vietnam, and Indonesia; these cases are frequently associated with direct contact with contaminated chickens.

Vietnam (2003–2005): Vietnam reported more than 90 human cases with a high death rate during the first wave of H5N1 outbreaks. The majority of infections were associated with contact with contaminated poultry, especially when food was being prepared.

Indonesia (2005–2007): Over 100 human cases of H5N1 were reported during a series of epidemics in Indonesia. A large number of these incidents were related to backyard poultry farming, in which owners lived adjacent to their chickens.

China (2005–2017): Over the years, the country has recorded isolated occurrences of H5N1 illnesses, which are frequently connected to exposure at live bird markets. Closing these marketplaces and putting sick birds to death have been major measures taken by the Chinese government to stop the spread.

Africa and the Middle East: New Hotspots

Aside from Asia, H5N1 has also afflicted the Middle East and Africa. While there have been fewer human cases in these places, there is still a risk of broad transmission.

Egypt (2006–2015): Egypt reported a high number of H5N1 cases in humans, with a large number of infections connected to domestic chicken. The risk was increased by cultural customs like housing chickens.

Nigeria (2006–2007): Nigeria saw H5N1 outbreaks in poultry, and while human cases were few, there was a chance of transmission because of the country's tight human–animal relationships.

Singular Incidents in North America and Europe

H5N1 infections, however rarer, have been documented in North America and Europe; these cases are usually linked to travelers returning from afflicted areas or direct contact with sick birds.

United Kingdom (2006): A individual who had visited an afflicted region in Asia was reported to have contracted H5N1.

Canada (2014): The country revealed the world's first H5N1 case in a tourist coming back from China, emphasizing the threat's worldwide scope.

Possibility of Spread from Person to Person and Pandemic Situations

The fact that H5N1 has the capacity to mutate and develop an effective method of human-to-human transmission is one of its most alarming characteristics. Even while there hasn't been a sustained human-to-human transmission yet, the high death rate linked to the virus makes the chance of one happening extremely dangerous.

Factors Influencing Transmission from Human to Human

Genetic Mutations: One of the characteristics of influenza viruses, particularly H5N1, is their capacity for genetic reassortment and mutation. Certain genetic alterations in the neuraminidase (NA) and hemagglutinin (HA) genes may improve the virus's capacity to attach to human receptors and promote human-to-human transmission.

Coinfection with Human Influenza Viruses: Reassortment, or the blending of genetic material from both viruses to produce a new strain possessing characteristics from both, may result from coinfection with human influenza viruses. This technique might produce an H5N1 virus that can spread effectively from person to person.

Environmental and Behavioral Factors: Dense populations, intimate relationships between people and animals, and lax biosecurity procedures can all contribute to the spread of H5N1. It is imperative that public health programs address these elements in order to prevent human infections.

Pandemic Situations

A number of different, mildly to moderately severe pandemic scenarios could occur if H5N1 were to have the capacity to spread among people.

Moderate Situation: Restricted Human-to-Human Transfer

In this case, H5N1 develops restricted human-to-human transmission capacities, resulting in epidemics that are restricted to certain areas and do not spread widely. Isolation, quarantine, and targeted immunization are examples of public health strategies that could successfully limit these outbreaks.

Moderate Situation: Dispersion in Regions

A moderate scenario entails H5N1 spreading more quickly among people, resulting in localized outbreaks that cause a considerable amount of morbidity and mortality. To stop the spread in this situation, major public health initiatives would be required, such as widespread use of antiviral drugs, travel restrictions, and mass vaccination campaigns.

Extreme Situation: Worldwide Pandemic

In the worst-case situation, H5N1 could mutate into a version that is easily transmitted among people, sparking a pandemic that affects the entire world. This scenario might have catastrophic effects, overwhelming healthcare systems and generating severe social and economic disruptions given the high fatality rate linked with H5N1.

Strategies for Response and Preparation

The risk can be reduced by using preparedness and response plans because H5N1 has the ability to spread widely and cause serious human illness.

Monitoring and Surveillance

For the early detection of H5N1 outbreaks in both animal and human populations, effective monitoring systems are crucial. This includes:

Animal Surveillance: Animal surveillance involves keeping an eye out for indications of H5N1 illness in wild bird populations, live bird markets, and poultry farms. Quick identification and reporting can aid in stopping the virus's spread.

Human Surveillance: Strengthening the lookout for H5N1 cases in humans, especially in high-risk populations including poultry workers and visitors returning from impacted areas. Early intervention depends on prompt diagnostic testing and reporting.

Strategies for Vaccination

A vital part of being ready for a pandemic is developing and storing H5N1 vaccinations. Despite the availability of candidate vaccines, issues with production capacity and guaranteeing that vaccinations correspond with circulating strains still exist.

Pre-pandemic vaccination: Immunizing high-risk individuals, such healthcare professionals and poultry workers, can add an extra layer of defense and lower the chance of contracting diseases from humans.

Pandemic Vaccination: It will be crucial to quickly distribute vaccines to the general public in the case of a pandemic. It is imperative to guarantee fair access to immunizations, especially in environments with limited resources.

Antiviral Drugs

Antiviral drugs like zanamivir and oseltamivir help shorten the duration and severity of influenza infections. It is essential to have these drugs in stock and make sure they are available in case of an outbreak.

Treatment: Giving infected people antiviral medication can lower their risk of illness and death. For efficacy, early administration is essential.

Prophylaxis: Antivirals can be used to stop epidemics in high-risk populations or in people who have already been exposed to the virus.

Measures of Public Health

In the lack of efficacious vaccinations, non-pharmaceutical interventions (NPIs) are essential for containing the H5N1 pandemic.

Isolation and Quarantine: To stop the spread of the infection, infected people should be kept apart from their contacts and kept in isolation. Strong public health infrastructure and transparent public communication are necessary for these actions.

Travel limitations: You can lessen the chance of bringing the virus into your country by putting in place travel limitations or screening procedures for visitors from impacted areas.

Community Engagement: In order to minimize transmission, it is crucial to work with communities to encourage preventive practices including washing your hands frequently, handling poultry carefully, and avoiding contact with sick birds.

The H5N1 avian flu has serious health consequences for humans since it can cause serious illness and high fatality rates. Although there is now little risk to the general public, it is important to be vigilant and prepared in case the virus mutates to allow for effective human-to-human transmission.

The threat posed by H5N1 can be reduced with the help of strong public health initiatives, vaccination plans, efficient surveillance, and antiviral stockpiling. A thorough response must also include public awareness and involvement to guarantee that people and communities are informed and ready to take the necessary precautions to safeguard others as well as themselves.

Maintaining preparedness and lowering the danger of a pandemic will need ongoing national and international study and collaboration as the H5N1 scenario develops. In order to effectively handle this intricate and persistent danger, public health organizations must build on the lessons garnered from previous outbreaks and their ongoing efforts.

Chapter 6

Development and Distribution of Vaccines

Synopsis of H5N1 Vaccine Investigation and Advancement

The Background of Influenza Vaccine History

The first inactivated influenza vaccine was developed in the 1940s, beginning a lengthy history of vaccine development against influenza viruses. Because influenza viruses are known to undergo significant genetic changes, seasonal flu vaccines must be updated on a regular basis. However, because of the features and epidemiology of avian influenza viruses like H5N1, developing vaccines against them poses special difficulties.

H5N1 Vaccine Development Challenges

There are various scientific and practical obstacles in the way of creating a vaccine against H5N1 avian flu:

Antigenic variety: There is a great deal of genetic and antigenic variety in H5N1 viruses. This means that vaccinations have to be carefully designed to target the strains that are circulating or are expected to circulate, necessitating ongoing surveillance and compositional modifications to the vaccine.

Production Time: Egg-based techniques are used to create traditional influenza vaccinations, which can take some time. In the event of a pandemic outbreak, this approach might not be quick enough, underscoring the necessity for alternate production techniques.

Adjuvants and Immunogenicity: Adding adjuvants—substances that stimulate the body's immunological response to the vaccine—to H5N1 vaccinations frequently improves their immunogenicity. One of the most important steps in developing vaccines is determining and testing appropriate adjuvants.

Safety and Efficacy: Extensive clinical trials are required to ensure the safety and efficacy of H5N1 vaccinations. These trials must show that the vaccine can elicit a robust immune response and protect against infection without generating notable side effects.

Research on H5N1 Vaccine Advances

Significant advancements have been made in the research of H5N1 vaccines over the last 20 years:

Pre-Pandemic vaccinations: A number of nations have developed and accumulated pre-pandemic H5N1 vaccinations. These vaccines are meant to be used in high-risk populations or during the initial stages of an outbreak. They are made to defend against certain H5N1 strains that have been discovered through surveillance activities.

Candidate Vaccines: Several vaccines are being developed and tested at different phases. These consist of recombinant, live attenuated, and inactivated vaccinations. In preclinical and clinical trials, a few potential candidates have demonstrated the capacity to trigger potent immune responses.

Creative Methods: Scientists are investigating new vaccine platforms and technologies, like COVID-19-like mRNA vaccines, viral vector vaccines, and nanoparticle-based vaccines. These creative methods could lead to more flexible and quick production.

Specifics of the Contract between CSL Seqirus and the CDC

Context and Justification

The world's top influenza vaccine producer, CSL Seqirus, entered into a deal with the Centers for Disease Control and Prevention (CDC) in late May to develop and stockpile H5N1 vaccinations. The United States is working to improve pandemic preparedness and guarantee a quick response in the event of an H5N1 epidemic. This strategic alliance is one aspect of that effort.

Goals of the Agreement

The agreement with CSL Seqirus seeks to accomplish a number of important goals:

Production of Vaccines: The main objective is to generate large quantities of an H5N1 vaccine that closely resembles the H5N1 virus that is already in circulation. This entails the formulation of completed vaccination doses as well as the large-scale manufacture of vaccine antigen.

Stockpiling: Making sure that there is a sizable supply of H5N1 vaccine doses on hand for quick distribution in the event of an outbreak. The goal of the stockpile is to safeguard vulnerable groups and offer first protection while the manufacture of more vaccines ramps up.

Increasing Capacity: Creating and preserving the ability to generate H5N1 vaccinations fast in the event of a pandemic danger. This involves making certain that the required labor force, raw resources, and infrastructure are available.

Important Elements of the Agreement

Vaccine Strain Selection: The antigenic properties of circulating H5N1 viruses and the surveillance data available today are used to determine which vaccine strain should be produced. The strain that is chosen needs to be typical of the viruses that are most likely to be harmful to humans.

Use of Adjuvants: The contract contains clauses pertaining to the use of adjuvants to improve the vaccine's immunogenicity. Adjuvants can increase the vaccine's effectiveness and possibly even allow for dose-sparing, which would allow for the vaccination of more people with a finite amount of vaccine.

Regulatory Compliance: Ensuring that the vaccine satisfies all safety, effectiveness, and quality requirements set forth by regulations. The process entails stringent examination and assessment by regulatory bodies like the Food and Drug Administration (FDA) in the United States.

Worldwide Vaccine Distribution and Stockpiling Strategies

The Value of a Worldwide Perspective

International cooperation is needed to address the worldwide concern of the H5N1 pandemic danger. Since influenza viruses have no geographical boundaries, an outbreak in one area can travel swiftly to adjacent areas. Thus, international vaccination distribution and storage plans are necessary for efficient pandemic preparedness and response.

Global Cooperation

World Health Organization (WHO): In order to coordinate worldwide influenza surveillance and response activities, the WHO is essential. In addition to keeping track of influenza viruses that are circulating, the WHO Global Influenza Surveillance and Response System (GISRS) offers recommendations for vaccination strain selection.

Global Vaccine Action Plan (GVAP): Supported by the World Health Assembly, the GVAP seeks to improve immunization programs globally and provide fair access to vaccines. It contains particular preparation methods for pandemic influenza.

International Partnerships: Information, resources, and knowledge can only be shared through partnerships between governments, international organizations, and the commercial sector. The Global Health Security Agenda (GHSA) and the Coalition for Epidemic Preparedness Innovations (CEPI) are two examples.

Methodical Stockpiling

National Stockpiles: As part of their pandemic preparedness strategies, several nations keep H5N1 vaccinations in reserve at the national level. The goal of these stockpiles is to give vital personnel and high-risk populations quick access to immunizations.

Regional Stockpiles: Coordinated vaccination distribution across several nations is made possible by regional stockpiles, which are overseen by institutions like the Asia-Pacific Economic Cooperation (APEC) and the European Centre for Disease Prevention and Control (ECDC).

Worldwide Vaccine Stockpile: The World Health Organization (WHO) keeps a worldwide supply of H5N1 vaccines, which is financed by donations and overseen by producers of the vaccines. This reserve is an essential tool for controlling epidemics in low- and middle-income nations.

Distribution Techniques

Prioritization: Based on risk assessments, vaccine delivery in the case of an outbreak must be arranged in order of priority. Immunizations are usually given priority to high-risk groups, such as healthcare personnel, poultry workers, and people with underlying medical disorders.

Logistics and Supply Chain Management: Strong supply chain management and logistics are necessary for effective distribution in order to guarantee that vaccinations reach the areas where they are most needed. This covers the cold chain infrastructure needed to store and distribute vaccines while maintaining their potency.

Equitable Access: A cornerstone of global health is guaranteeing immunizations to all people equally. Disparities in vaccine distribution and availability need to be addressed, especially in environments with low resources.

Frameworks for Allocating Vaccines

WHO Allocation system: To direct the delivery of pandemic influenza vaccines, the WHO has created an allocation system. Allocation is prioritized using this paradigm according to the ability to limit transmission, danger of exposure, and influence on public health.

COVAX Facility: The COVAX Facility offers a model for fair vaccination distribution, notwithstanding its primary focus on COVID-19. It seeks to guarantee that during a pandemic, vaccines are available to all nations, irrespective of their economic status.

National Allocation Plans: Governing bodies are required to create national allocation plans that specify the distribution of vaccines inside their borders. These strategies ought to be grounded on logistical considerations, ethical standards, and epidemiological facts.

One of the most important aspects of worldwide pandemic preparedness and response is the creation and dissemination of H5N1 vaccinations. Despite tremendous advancements in our knowledge of the virus and the creation of potent vaccines, obstacles

still lie in the way of guaranteeing that these life-saving treatments are quickly and fairly accessible.
The CDC's agreement with CSL Seqirus is a proactive move that will improve the country's readiness for an H5N1 outbreak. Building production capacity and securing a vaccination stockpile have improved the CDC's ability to safeguard the public in an emergency.
Coordinated actions by governments, the business sector, and international organizations are necessary to combat the H5N1 danger on a worldwide scale. Mitigating the consequences of a potential pandemic requires strategic stockpiling, efficient distribution plans, and fair access to vaccines.

Sustaining preparedness and preserving public health will require constant research, surveillance, and cooperation as the field of infectious diseases changes. Future plans will be informed by the knowledge gained from previous and ongoing efforts in vaccine development and distribution, making us more equipped to handle the difficulties posed by newly emerging infectious diseases.

Chapter 7

Farm Safety Precautions

Optimal Methods for Biosecurity in Farms that Raised Dairy and Poultry

The term "biosecurity" describes the policies and procedures put in place to keep pathogens from entering and spreading throughout a farm. To stop H5N1 avian influenza outbreaks and other infectious disease outbreaks in dairy and poultry farms, effective biosecurity is essential.

Important Farm Biosecurity Elements

Control around the Periphery:

Fencing: Secure fence is needed to keep wild animals, which can spread illness, out of the farm and to prevent unwanted access.

Controlled Access Points: Specific entry and departure locations equipped with security features to regulate the flow of people, cars, and equipment.

Management of Animal Health

Quarantine: Separate newly arrived or returned animals for a while (often 30 days) before reintegrating them into the flock or herd.

Health Monitoring: Consistent examinations for symptoms of disease in animals, along with prompt isolation and veterinary attention for any that do.

Visitor and Employee Etiquette

Visitor Restrictions: Only needed staff should be able to access the farm. Make guests sign in and adhere to biosecurity procedures.

Protective Clothing: Give visitors and employees access to and mandate the usage of hygienic, farm-appropriate apparel and footwear. The best boots and coveralls are disposable ones.

Water and Feed Management

Safe Feed Sources: To avoid contamination, obtain feed from reliable vendors and store it in a clean, safe space.

Clean Water: Ensure that animals have access to uncontaminated water that isn't possibly harboring germs.

Control of Rodents and Pests

Rodent Proofing: Use traps or baits along with entry point sealing to keep rodents out of structures.

Pest Control Programs: Regular pest management procedures are necessary to lessen the number of insects and other pests that may spread disease.

Sanitizing and disinfecting

Frequent Cleaning: To preserve hygiene, clean the feeding apparatus, animal housing, and other farm spaces on a regular basis.

Disinfection Procedures: To disinfect surfaces, tools, and clothing that may have come into contact with microorganisms, use approved disinfectants and adhere to the prescribed procedures.

Procedures for Disinfection and Equipment Administration

An essential part of biosecurity on farms is disinfection, which aims to eliminate or deactivate microorganisms to stop the spread of illnesses like H5N1. Upholding farm hygiene requires appropriate equipment management and effective disinfection procedures.

Procedures for Disinfection

Selecting the Appropriate Disinfectant

Efficacy: Choose disinfectants with a track record of success against dangerous infections, such as H5N1.

Safety: Verify that using the disinfectant near animals is safe and won't endanger agricultural laborers.

Getting Ready and Using It

Surface Cleaning: Prior to using disinfectants, wash surfaces with soap and water to get rid of any organic matter. Disinfectants may be less effective when organic stuff is present.

Dilution: To guarantee efficacy, dilute disinfectants according to the manufacturer's recommendations.

Application: To guarantee complete coverage of all surfaces, apply disinfectants using the proper techniques, such as spraying, fogging, or soaking.

Time of Contact

Respect for Guidelines: To effectively kill or inactivate microorganisms, make sure disinfectants are applied to surfaces for the full amount of time advised.

Frequently

Regular Intervals: Make sure to schedule regular disinfection tasks, especially in high-risk areas including equipment rooms, feed storage facilities, and animal housing.

Equipment Oversight: Specific Equipment

Farm-Specific Tools: To avoid cross-contamination, use specific tools and equipment for each section of the farm.

Color-Coding: To differentiate equipment for different sections, such as feeding, cleaning, and medical supplies, use a color-coding scheme.

Cleaning Techniques

Routine Cleaning: Whenever possible, wipe off equipment after using it, especially if it has come into touch with organic materials or animals.

Disinfection: After cleaning, disinfect the equipment, being careful to get into tight spaces where bacteria may hide.

Maintenance and Storage

Appropriate Storage: To avoid contamination and damage, keep equipment stored in approved locations. Make sure storage spaces are dry and clean.

Maintenance: To keep equipment in good operating order, give it regular inspections and maintenance. Replace worn-out or broken equipment right away.

Case Studies of Effective Containment Initiatives

The effective use of disinfection procedures and biosecurity measures in limiting H5N1 epidemics on farms is demonstrated by a number of case studies. These illustrations offer insightful information about successful tactics and industry best practices.

Case Study 1: H5N1 Outbreak Contained on a Vietnamese Poultry Farm

Vietnam has seen multiple H5N1 outbreaks because of its extensive chicken sector and tight relationships between humans and animals. In one well-known instance, a medium-sized chicken farm used strict biosecurity protocols to successfully contain an outbreak.

Quick Reaction

Early Detection: A strong surveillance system on the farm enabled the early identification of the virus in a small number of birds.

Quick Isolation: To stop the infection from spreading, the afflicted area was roped off and infected birds were isolated right away.

Quarantine and Disinfection

Thorough Cleaning and Disinfection: The farm was thoroughly cleaned and disinfected. Approved disinfectants were used to treat all surfaces and equipment.

Measures of Quarantine: The farm was put under quarantine, limiting the movement of workers, equipment, and animals until the outbreak was controlled.

Coordinating and Communicating

Local Authorities: To guarantee adherence to biosecurity rules and obtain advise, the farm management collaborated with local veterinary authorities.

Community Involvement: In order to spread knowledge about the outbreak and promote the reporting of sick birds, the farm held community meetings.

Result

Effective Containment: The infection was effectively contained with little to no spread, highlighting the value of prompt action and strict biosecurity protocols.

Case Study 2: Colorado, USA Dairy Herd Prevention of H5N1 Transmission

The H5N1 virus posed a threat to a Colorado dairy farm, which successfully contained the virus and stopped it from spreading.

Improved Safety and Health

Visitor Restrictions: The farm only allowed access to necessary staff members and mandated that all guests adhere to stringent biosecurity procedures, which included washing their clothes and shoes.

Protective Measures: Personal protective equipment (PPE) and training on correct handling and cleaning techniques were given to the staff.

Procedures for Disinfection

Daily Cleaning: Using potent disinfectants, the housing for the animals and the equipment used for milking were cleaned and sanitized every day.

Vehicle Disinfection: To stop the virus from being introduced or spreading, all vehicles coming into and going out of the farm were cleaned.

Monitoring Health

Frequent Testing: The farm tested the animals frequently to look for indications of infection. Animals exhibiting symptoms were evaluated and isolated right away.

Veterinary Support: To keep an eye on the herd's health and take extra action when necessary, the farm collaborated closely with veterinarians.

Result

No Spread Found: The farm's improved biosecurity and disinfection protocols stopped the H5N1 virus from spreading, safeguarding the animals as well as the larger population.

Case Study 3: Effective Management of H5N1 in an Egyptian Poultry Farm

H5N1 epidemics have also presented difficulties for Egypt, especially in smallholder poultry farms. One noteworthy instance of confinement that worked well was a tiny farm that followed strict biosecurity procedures.

Local Education

Awareness Campaigns: To increase public knowledge about H5N1 and the value of biosecurity, the farm took part in neighborhood education initiatives.

Training: Farmers received instruction on biosecurity best practices, such as appropriate cleaning and disinfection methods.

Biosecurity Procedures

Controlled Access: There were safeguards in place to keep unwanted people and cars out of the farm. Access was restricted.

Protective Clothes: Before and after accessing the regions used for housing poultry, farmers cleaned their protective clothes and shoes.

Monitoring Diseases

Frequent Monitoring: The farm established a routine for keeping an eye out for illness indicators. Any questionable symptoms were promptly reported to the veterinary authorities.

Testing: As soon as a case is detected, it should be tested to rule out H5N1 and start the containment process.

Result

Effective Control: The outbreak was successfully contained and did not spread, underscoring the need of public awareness campaigns and strict adherence to biosecurity guidelines.

In order to stop the spread of H5N1 avian influenza and other infectious diseases, farmers must take preventative measures. Effective disease prevention is based on best practices in biosecurity, which include perimeter control, pest control, cleaning and disinfection, visitor protocols, feed and water management, and animal health management.

In order to ensure that pathogens are properly eliminated or rendered inactive and that farm equipment does not serve as a vector for the spread of disease, disinfection procedures and equipment management are essential elements of biosecurity. The effective application of these strategies in case studies from Egypt, the United States, and Vietnam highlights the significance of adhering to best practices, including the community, and acting quickly in order to prevent and contain H5N1 outbreaks.

In order to maintain strong biosecurity protocols and safeguard the health of both humans and animals, the global farming community must continue to collaborate, educate, and train its workforce on new and emerging infectious illnesses, such as avian influenza.

Chapter 8

Global Initiatives and Policies

International Organizations' Function

Global efforts to prevent, track, and treat infectious diseases like H5N1 avian influenza are mostly coordinated by international agencies. The Food and Agriculture Organization of the United Nations (FAO) and the World Health Organization (WHO) are two important institutions participating in these initiatives.

World Health Organization (WHO)

Monitoring and Surveillance: To track the spread of H5N1 and other infectious illnesses, the WHO organizes international surveillance operations. The World Health Organization (WHO) gathers and evaluates data from member states via its Global Influenza Surveillance and Response System

(GISRS) in order to spot new risks and guide response plans.

Advice and Suggestions: The World Health Organization offers member governments advice and suggestions for illness prevention, control strategies, and treatment regimens. These recommendations aid in standardizing methods for managing epidemics and are founded on best practices and scientific data.

Building Capacity: The World Health Organization (WHO) offers training, technical support, and resources to enhance laboratory infrastructure, emergency response capabilities, and healthcare systems in its member states.

The United Nations Food and Agriculture Organization (FAO)

Animal Health Surveillance: Through its Emergency Prevention System for Animal Health (EMPRES), the FAO is dedicated to keeping an eye on and managing animal illnesses, such as avian influenza. The FAO enhances animal population observation, early identification, and response to outbreaks by collaborating with member nations.

Support for Livelihoods: The Food and Agriculture Organization (FAO) not only fights disease but also helps farmers and rural communities impacted by outbreaks. This helps to reduce the financial burden of disease control efforts and avert crises related to food security.

One Health Approach: The Food and Agriculture Organization (FAO) advocates for the One Health approach, which acknowledges the interdependence of environmental, animal, and human health. The FAO seeks to lessen the emergence and spread of zoonotic illnesses like H5N1 by addressing health concerns at the interface between humans, animals, and the environment.

The Impact of Import/Export Limitations on the Transmission of Diseases

Countries frequently impose import and export prohibitions in an effort to stop the entry or spread of contagious viruses such as the H5N1 avian influenza. These actions may be successful in reducing the spread of disease, but they may also have negative social and economic effects.

Import/Export Limitations' Effects

Trade Disruptions: Trade in agricultural products, particularly chicken and poultry products, can be affected by import and export restrictions. Export prohibitions enforced by impacted nations may result in decreased market access for manufacturers and exporters, which may cause financial losses and trade conflicts.

Disruptions to the Supply Chain: Limitations on the flow of people and things can cause supply networks to break down, which can result in a shortage of necessities. This may have a domino effect on sectors that depend on agricultural inputs, such as food processing and manufacturing.

Price Volatility: In agricultural markets, disturbances in trade and the supply chain can cause price volatility that impacts both producers and consumers. Reduced availability of chicken and poultry products can lead to price surges, while decreased demand for exports can cause prices to fall in home markets.

Livelihoods and Food Security: Farmers' livelihoods, particularly those of smallholder farmers who depend on foreign markets, can be significantly impacted by import/export restrictions. Decreased export revenue has the potential to threaten rural populations' food security and means of subsistence, hence escalating poverty and inequality.

Joint Surveillance and Quick Reaction Initiatives

Effective surveillance and prompt reaction to H5N1 avian influenza epidemics require cooperation between nations and international organizations.

Through the exchange of knowledge, resources, and skills, nations can enhance their ability to identify, manage, and lessen the consequences of infectious illnesses.

Networks of Multilateral Surveillance

Global Influenza Surveillance and Response System (GISRS): GISRS, or the Global Influenza Surveillance and Response System, GISRS, a global network of labs and surveillance centers overseen by the WHO, is in charge of keeping an eye on circulating influenza viruses, including H5N1. To help with pandemic preparedness and vaccine strain selection, member nations provide samples and data.

OFFLU: The OIE/FAO Network of Expertise on Avian Influenza (OFFLU) is a cooperative network that unites specialists in animal health to offer advice and technical support on the prevention and control of avian influenza. To keep an eye on influenza viruses in animal populations, OFFLU collaborates closely with GISRS.

Information Sharing and Regional Collaboration

Regional Networks: Cooperation and information exchange between neighboring nations are facilitated by regional institutions like the African Union Inter-African Bureau for Animal Resources (AU-IBAR) and the European Centre for Disease Prevention and Control (ECDC). At the regional level, these networks improve surveillance and response capacities.

Collaborative Training and Exercises: To evaluate their readiness and capacity to respond to health crises, nations engage in cooperative training exercises and simulations. These activities improve public health professionals' cooperation, coordination, and communication.

Public-Private Collaborations

Engagement of the Private Sector: Public-private alliances are essential to the monitoring and management of disease outbreaks. Research, development, and distribution of vaccines and other interventions are supported by the knowledge, assets, and financial contributions of pharmaceutical corporations, vaccine producers, and agribusinesses.

Research Collaboration: To get a deeper understanding of the epidemiology, dynamics of transmission, and pathophysiology of H5N1 avian influenza, academic institutions, research organizations, and industry partners work together on research initiatives. These collaborations stimulate creativity and contribute to the development of evidence-based disease control plans.

In order to combat the worldwide threat posed by H5N1 avian influenza and other infectious diseases, international cooperation and regulations are needed. Countries can cooperate through agencies such as the WHO and FAO to improve surveillance, exchange data, and plan reaction actions in order to stop and manage epidemics.

Restrictions on imports and exports can stop the spread of disease, but they can also have unforeseen effects like disrupting trade and hurting the economy. It is crucial to strike a balance between socioeconomic factors and public health goals, and to make sure that any interventions are transparent, evidence-based, and risk-appropriate.

Early detection and containment of H5N1 outbreaks depend heavily on cooperative efforts in surveillance and quick reaction. Cooperating across industries and national boundaries can help nations improve their capacity for reaction and readiness, lessen the effects of infectious diseases, and protect food security and public health.

Chapter 9

H5N1 Medical Treatments

Bird flu, also known as avian influenza A (H5N1) virus, is a serious health risk to humans because of its high fatality rate and ability to induce severe respiratory illnesses. Medical interventions are essential in managing infected persons and mitigating the severity of sickness, even if prevention through vaccination and biosecurity measures continues to be the key strategy to stop the spread of H5N1. An overview of H5N1 medical treatments, including antiviral drugs, treatment regimens, and potential future developments for novel medicines, is given in this chapter.

An Overview of Medicines That Work

Tamiflu (oseltamivir)

Oseltamivir is a neuraminidase inhibitor that lessens viral reproduction and spread inside the body by preventing freshly produced virus particles from being released from infected cells.

Clinical research has demonstrated that administering oseltamivir early on to patients infected with H5N1 influenza can shorten the duration and lessen the intensity of symptoms.

Nonetheless, the appearance of H5N1 virus strains resistant to oseltamivir highlights the significance of ongoing monitoring and the creation of substitute therapeutic approaches.

Zanamivir (Swine Flu)

Another neuraminidase inhibitor that prevents the growth of viruses is zanamivir; it works by specifically inhibiting the neuraminidase enzyme found on the surface of influenza viruses

Similar to oseltamivir, zanamivir has also been demonstrated to be useful in lowering the intensity and length of H5N1 influenza symptoms in individuals.

For individuals who are unable to tolerate oral treatment, zanamivir is available as an inhalation powder that is delivered orally.

Treatment Plans for Those Who Are Infected

Early Treatment Commencement

For patients with H5N1 influenza, prompt antiviral medication initiation is essential to optimizing therapeutic efficacy and lowering the risk of sequelae.

Antiviral therapy should be started as soon as symptoms appear, ideally within 48 hours, according to current treatment guidelines, however benefits might still be shown if treatment is started later in the course of the illness.

Treatment Duration

The length of time that antiviral therapy is advised for H5N1 influenza varies based on the patient's unique circumstances and the severity of their sickness.

Longer durations of therapy may be considered for patients with severe or complex illnesses. Generally, treatment is administered for a minimum of five days or until symptoms disappear.

Supplementary Treatments

Supportive treatment interventions including extra oxygen, mechanical ventilation, and hydration and electrolyte replenishment may be required in addition to antiviral drugs to address complications of severe respiratory illness linked to H5N1 influenza.

While they are not usually advised, corticosteroids are occasionally used to treat severe influenza-related pneumonia. However, their effectiveness is still debatable.

Prospects for New Antiviral Drugs in the Future

Inhibitors of Neuraminidase of the Next Generation

There are now a number of experimental neuraminidase inhibitors being developed that have greater potency and broader range efficacy against influenza viruses.

Patients with H5N1 influenza may have better treatment choices thanks to these next-generation antiviral treatments, which may be able to circumvent resistance mechanisms seen with current medications.

Treatments Aimed at the Host

Another promising strategy for treating H5N1 influenza is host-targeted therapy, which alters the host immune response to influenza infection.

By minimizing tissue damage and the overwhelming inflammatory response that occur with a severe influenza infection, these treatments hope to improve clinical outcomes and lower death.

All-natural Antibodies

Potential therapeutic medicines under research are monoclonal antibodies that target particular viral antigens or host factors implicated in the development of H5N1 influenza.

In high-risk populations including healthcare professionals and people with underlying medical issues, these monoclonal antibodies may offer quick and focused immune protection against H5N1 infection.

When it comes to managing H5N1 influenza, medical interventions are essential, especially when it comes to lessening the intensity of the sickness and shielding those who are infected from complications. When started early in the course of sickness, currently available antiviral drugs like oseltamivir and zanamivir are helpful in suppressing viral replication and alleviating symptoms. Drug-resistant viruses, on the other hand, emphasize the necessity of ongoing research and development of novel antiviral medications with increased potency and efficacy against H5N1 influenza.

Promising options for improving clinical outcomes and lowering mortality in patients with H5N1 influenza include host-targeted treatments, monoclonal antibodies, and next-generation neuraminidase inhibitors. In order to combat the growing threat of avian influenza and safeguard public health worldwide, sustained investment in research and development as well as strong surveillance and monitoring operations will be crucial.

Chapter 10

Safety Advice and Public Health Guidelines

In order to stop the spread of infectious diseases like the H5N1 avian influenza and shield people and communities from illness, public health recommendations and safety measures are crucial. An summary of individual personal preventive measures, travel advisories and guidance for high-risk places, and useful advice for minimizing viral exposure are all included in this chapter.

Individuals' Personal Protective Measures

Hand Sanitization

Hands should be periodically washed for at least 20 seconds with soap and water, especially after sneezing or coughing, before eating, and after coming into touch with animals or their surroundings.

Use alcohol-based hand sanitizer with at least 60% alcohol if soap and water are not available.

Breathing Hygiene

When you sneeze or cough, cover your mouth and nose with your elbow or a tissue. After using a tissue, quickly dispose of it and wash your hands.

To stop the spread of germs, avoid touching your face, especially your lips, nose, and eyes.

The application of personal protective equipment (PPE)

When caring for sick animals or being in close proximity to someone who may be infected with H5N1, wear a properly fitting medical mask or respirator (such as the N95).

Handle potentially contaminated goods or surfaces with gloves on and other personal protective equipment (PPE) as needed.

Social Distancing

Keep a minimum of 6 feet (2 meters) between you and other people, especially if they are sneezing, coughing, or otherwise exhibiting symptoms of a respiratory disease.

Steer clear of crowded areas and close contact with ill people since they may increase the chance of transmission.

Travel Warnings and Recommendations for Areas at High Risk

Limitations on Travel

Keep abreast of travel warnings and limitations issued by international and national health authorities, especially for regions where H5N1 avian influenza outbreaks are occurring.

Delay non-essential travel to high-risk locations if you are more susceptible to serious influenza-related illnesses or consequences.

Safety Measures for Travel

If you must travel, take measures to reduce your chance of contracting H5N1 influenza. These include avoiding sick people's company and maintaining proper hand and respiratory cleanliness. Recognize the local health rules and regulations pertaining to traveler testing procedures, requirements for quarantines, and the use of personal protective equipment.

Following Travel, Health Monitoring

After returning from a trip to a high-risk area, keep a watchful eye on your health for signs of respiratory illness, such as fever, coughing, and breathing difficulties, for at least 14 days.

If you have symptoms that could be related to the H5N1 influenza, get medical help as soon as possible, and let the medical staff know about any previous travel experiences.

Useful Advice for Lowering Your Virus Exposure

Steer clear of sick animals

Steer clear of birds, chickens, and other animals that seem ill or that have passed away unexpectedly, particularly in regions where H5N1 influenza has been identified.

To lower the risk of infection, avoid handling or eating any raw or undercooked poultry items, including eggs.

Handle food safely

165°F (74°C) is the minimum internal temperature at which chicken and eggs must be cooked to eliminate any potentially dangerous bacteria or viruses.

To avoid contaminating raw poultry items with other foods, use different utensils and cutting boards.

Keep Your Living Areas Tidy

Maintain clean, well-ventilated living areas and clean and sanitize high-touch surfaces like light switches, countertops, and doorknobs on a regular basis.

Get rid of waste, even animal dung, in a way that reduces the chance of contaminating the environment.

Remain Up to Date

Get up to date on the most recent information about H5N1 avian influenza from reliable sources like the CDC, WHO, and national health authorities.

In order to prevent disease in your community and yourself, abide by the local public health recommendations and guidelines.

In order to stop the spread of H5N1 avian influenza and lower the risk of infection in both individuals and communities, public health recommendations and safety measures are crucial. The use of personal protective equipment (PPE), social distance, hand and respiratory cleanliness, and other preventive measures are vital in reducing the spread of the virus.

Travel warnings and instructions for high-risk areas assist people in making well-informed travel decisions and implementing the necessary safety measures to reduce their chance of contracting the H5N1 influenza. Infection can also be avoided by following sensible advice on lowering exposure to the virus, such as avoiding sick animals and handling food safely.

Individuals can help prevent the spread of infectious diseases worldwide and safeguard themselves and others from H5N1 avian influenza by adhering to these recommendations and implementing them into regular activities.

Chapter 11

Observation and Readiness

The H5N1 form of avian influenza, in particular, continues to be a hazard to public health around the world. Vigilant surveillance and preparatory measures are necessary to stop and lessen future outbreaks. This chapter examines methods for enhancing international preparedness for avian influenza as well as the function of surveillance systems, early warning networks, technology, and data analysis in disease monitoring.

Early Warning Systems and Surveillance Systems

Worldwide Surveillance Systems

Global surveillance activities are coordinated by international agencies like the Food and Agriculture Organization of the United Nations (FAO) and the World Health Organization (WHO) to track the spread of avian influenza strains, including H5N1.

These surveillance networks gather and examine information on influenza viruses that are circulating in human and animal populations in order to improve response plans and provide early warning of new dangers.

National Watchdog Initiatives

Numerous nations have instituted nationwide surveillance initiatives to track cases of avian influenza in domestic and wild bird populations, as well as in susceptible humans.

These efforts include surveillance of human cases of avian influenza, routine testing of poultry and wild birds for influenza viruses, and continuous monitoring of bird populations.

Early Alerting Mechanisms

Predictive modeling, laboratory testing, and epidemiological data are used by early warning systems to identify and monitor avian influenza epidemics in real time.

Early warning systems for avian influenza must include speedy laboratory testing, informational dissemination, and rapid reporting of suspected cases.

The function of data analysis and technology

Genomics-Based Monitoring

The monitoring of avian influenza viruses has been transformed by advances in genome sequencing technology, which enable scientists to examine the genetic composition of different viral strains and monitor their temporal progression.

The development of vaccines and control strategies is guided by the important insights that genomic surveillance offers into the origins, dynamics of transmission, and genetic diversity of avian influenza viruses.

Analytics for Big Data

Large volumes of epidemiological, environmental, and molecular data can be integrated and analyzed using big data analytics to find patterns, trends, and risk factors related to avian influenza outbreaks.

To predict the emergence and spread of avian influenza strains and guide targeted actions, surveillance data can be subjected to machine learning algorithms and predictive modeling techniques.

Technologies for Remote Sensing

Geographic information systems (GIS) and satellite images are two examples of remote sensing technologies that offer useful tools for tracking environmental conditions that affect the spread of avian influenza.

These tools are capable of tracking the migration patterns of migratory bird populations, identifying high-risk areas for avian influenza outbreaks, and evaluating the effects of environmental changes on disease transmission.

Techniques to Boost International Readiness

Improved Cooperation and Exchange of Information

To increase worldwide readiness for avian influenza, cooperation and information exchange between nations, international organizations, and research institutes must be strengthened.

Improved data exchange, collaborative research projects, and well-coordinated reaction activities can aid in the early detection of new risks and enable the quick control of epidemics.

Developing Capabilities and Training

Enhancing preparedness and response capacities requires funding capacity building and training programs for public health experts, veterinarians, and healthcare personnel.

To promote a coordinated and efficient response to avian influenza outbreaks, training initiatives should concentrate on surveillance techniques, laboratory diagnostics, outbreak investigation, and risk communication.

One Method for Health

To effectively address the complex difficulties posed by avian influenza, a One Health strategy that acknowledges the interconnection of human, animal, and environmental health is needed.

At the intersection of humans, animals, and the environment, cooperative efforts between the human and animal health sectors and environmental agencies can enhance the early identification, prevention, and control of avian influenza epidemics.

Global efforts to stop the spread of avian influenza and lessen its effects on human and animal health are largely focused on preparation and monitoring. The identification and tracking of avian influenza epidemics is greatly aided by surveillance systems, early warning networks, and sophisticated technology. Strategies for enhancing global preparedness center on cooperation, capacity building, and a One Health philosophy.

Through fortifying surveillance capacities, utilizing technology and data analysis, and augmenting international cooperation, the international community can more effectively predict, ready for, and address any risks associated with avian influenza in the future. To protect public health and stop the spread of this deadly infectious illness, more funding must be allocated to monitoring and preparedness initiatives.

Chapter 12

Research and Innovations

In the fight against avian influenza, research and innovation are key factors as they propel developments in immunology, virology, vaccine science, and antiviral therapy. This chapter examines state-of-the-art research in these fields, focuses on developments in vaccine and antiviral therapy, and addresses future directions and possible scientific advances.

Cutting-Edge Immunology and Virology Research

Research on Genomes

Avian influenza virus genome sequencing offers important new information about the pathogenicity, evolution, and dynamics of viral transmission.

Scientists examine the genetic composition of virus strains, pinpoint genetic factors that contribute to virulence, and monitor the appearance of new variations using cutting-edge sequencing methods.

Host-Virus Dynamics

The processes by which avian influenza viruses multiply, infect, and elude host immune responses are clarified by studies of host-virus interactions.

Researchers look on immune evasion tactics used by the virus, host immunological pathways involved in antiviral defense, and host variables that affect susceptibility to infection.

Research on Immunology

The main goals of immunological research are to improve host immunity and comprehend the immune response to avian influenza infection.

Research delves into the function of both innate and adaptive immune responses in preventing avian influenza, along with the advancement of innovative immunotherapeutic strategies such immune modulators and monoclonal antibodies.

Advances in Antiviral Therapies and Vaccine Technology: Next-Generation Vaccines

The creation of vaccines with enhanced immunogenicity and efficacy can now be accomplished thanks to developments in vaccine technology, such as recombinant DNA technology and viral vector platforms.

New vaccine candidates provide broad-spectrum defense against a variety of influenza strains, including H5N1, by focusing on conserved influenza virus epitopes.

Adjuvants and Methods of Delivery

The immunogenicity and effectiveness of influenza vaccinations are improved by adjuvants and innovative delivery methods, allowing for dosage sparing and quick induction of protective immunity.

Targeted and needle-free vaccination methods are provided by mucosal vaccination tactics and platforms for vaccine delivery powered by nanotechnology.

Antiviral Treatments

Treatments against avian influenza that target viral replication and dissemination by blocking host cell entry factors and viral enzymes are known as antiviral treatments.

Broad-spectrum antivirals and host-targeted therapies are examples of next-generation antiviral medications that are being developed. These medications have improved potency, specificity, and resistance profiles.

Prospective Future Paths and Significant Advancements

All-around Influenza Vaccinations

Development of universal influenza vaccines that can offer cross-protective, long-lasting immunity against a variety of influenza strains, including H5N1, is the main goal of research activities.

Prospective universal vaccines aim to stimulate widespread and long-lasting immune responses against several influenza subtypes by targeting conserved epitopes of the influenza virus.

Immune-Systemic Treatments

Immunotherapeutic strategies, including immune checkpoint inhibitors and monoclonal antibodies, show promise in the management of severe influenza infections, particularly H5N1.

These treatments can reduce inflammation, improve viral clearance, and improve clinical outcomes in patients with severe respiratory illnesses by modifying the host immune response.

Monitoring of Antiviral Resistance

To track the establishment of drug-resistant strains and guide treatment decisions, ongoing observation of antiviral resistance in avian influenza viruses is crucial.

The main focus of research is on creating quick diagnostic assays and surveillance instruments to find mutations causing antiviral resistance in environmental and clinical samples.

Improvements in avian influenza prevention, diagnosis, and treatment are being driven by research and innovation. Innovative studies in immunology and virology shed light on host immune responses and viral pathogenesis, which helps create new vaccines and antiviral treatments. Promising strategies for battling avian influenza and lessening its worldwide impact are provided by advancements in vaccine technology and antiviral therapies.

Research will go in new directions in an effort to address outstanding issues including the creation of vaccinations against influenza that are universally available and the tracking of antiviral resistance. Through leveraging scientific innovation and teamwork, scientists can advance the battle against avian influenza and eventually safeguard the health of people and animals globally.

Chapter 13

Takeaways and Future Directions

In light of the persistent threat posed by avian influenza, namely the H5N1 strain, it is imperative to consider previous outbreaks, reactions, and lessons learned in order to choose the best course of action. This chapter offers an overview of the most important discoveries and their consequences, offers insights into previous outbreaks and reactions, and presents a plan for upcoming public health initiatives and research goals.

A synopsis of the main conclusions and their implications

Low Prior Immunization

The American populace appears to have little to no pre-existing immunity to H5N1 avian influenza, according to preliminary results from CDC serology investigations.

This emphasizes how crucial it is to have continuous surveillance, identify infections early, and act quickly to stop the spread of potentially pandemic-causing avian influenza viruses.

Persistent Epidemics in Agricultural Communities

Current outbreaks in poultry flocks and dairy herds draw attention to the persistent danger of animal-to-human avian influenza transmission, especially in rural areas.

Improving biosecurity protocols, monitoring schemes, and immunization plans are critical to managing epidemics and safeguarding the health of people and animals alike.

Effects on Human Health

Even though there is still little chance of H5N1 spreading from person to person, occasional human infections highlight the serious respiratory problems and even death that can result from avian influenza.

Effective public health measures, early diagnosis, and vigilance are essential for averting human outbreaks and reducing their effects on susceptible groups.

Thoughts on Previous Epidemics and Reactions

Past Pandemics

The destructive effects of influenza viruses on human health and society are brought to light by thinking back on previous influenza pandemics, such as the 1918 Spanish flu and the more recent H1N1 pandemic.

The necessity of early discovery, quick reaction, and international cooperation are among the lessons learnt from previous pandemics that guide current avian influenza preparedness initiatives.

Global Reaction Initiatives

The ability of governments, public health organizations, and international organizations to respond to and prepare for avian influenza has improved dramatically.

The ability to detect, prevent, and control avian influenza outbreaks has been increased by the creation of surveillance networks, vaccine stockpile initiatives, and research collaborations.

Prospects for Upcoming Public Health Initiatives and Research Focus Areas

An Integrated One Health Approach

Adopting a One Health paradigm that incorporates animal, human, and environmental health is crucial to tackling the multifaceted problems caused by avian influenza.

Together with environmental organizations, the human and animal health sectors can enhance surveillance, early disease detection, and response to zoonotic disease hazards through coordinated activities.

Cutting-edge technologies and research

The field of avian influenza prevention and control must advance through funding cutting-edge research and technologies, such as next-generation vaccines, antiviral treatments, and predictive modeling techniques.

Virus transmission patterns can be better understood and targeted interventions can be directed by utilizing the capabilities of big data analytics, genomic sequencing, and remote sensing technologies.

Building Capacity and Working Together

Enhancing global preparedness and response capabilities requires stepping up capacity building initiatives and encouraging cooperation among researchers, medical professionals, veterinary professionals, and public health experts.

The interchange of knowledge and resources to address the threat of emerging infectious diseases can be facilitated by training programs, knowledge-sharing websites, and cross-sectoral collaborations.

In conclusion, the worldwide response to avian influenza must be guided by reflection on past experiences, lessons learned, and future challenges. We may better prepare for and reduce the effects of avian influenza on human and animal health by reviewing important results, thinking back on previous outbreaks and responses, and laying out a vision for future public health initiatives and research objectives. We can work together, be innovative, and stay true to the One Health tenets to create a safer, healthier future for everybody.

Appendices

Appendix A: Terminology Glossary

Influenza A viruses are the source of Avian Influenza (AI), a highly contagious viral illness that affects birds. The severity of avian influenza can range from mild to severe, with some strains—like H5N1—posing a serious risk to both human health and poultry.

H5N1: An influenza subtype in birds a virus that can infect humans and other mammals but mainly affects birds. The high mortality rate of H5N1 influenza in birds and the occasional instance of severe respiratory illness in people are its defining characteristics.

Serology: Serology is the scientific study of physiological fluids, such as serum, with an emphasis on the immune system and the identification of antibodies that are directed against certain infections, like the H5N1 influenza virus.

Neuraminidase Inhibitors: Antiviral drugs that prevent the influenza virus's surface-bound protein, neuraminidase, from doing its job. Neuraminidase inhibitors are used to treat and prevent influenza infections, particularly H5N1. Examples of these inhibitors are zanamivir and oseltamivir.

One Health: An interdisciplinary strategy that acknowledges the connections between the health of people, animals, and the environment in order to address complicated health challenges. The One Health strategy places a strong emphasis on working together across all sectors to improve everyone's health.

Appendix B: List of References and Resources

Centers for Disease Control and Prevention (CDC): The CDC's official website with data on surveillance reports, public health recommendations, and avian influenza. Link

World Health Organization (WHO): Dedicated website with information on avian influenza, including technical resources, outbreak updates, and data from worldwide surveillance. Link

The Food and Agriculture Organization of the United Nations (FAO) maintains an avian influenza portal with news, publications, and recommendations for controlling and preventing the disease in animals. Link

Research Studies in Science

Donis RO and Smith GJ (2014) updated the nomenclature in response to the evolution of the avian influenza A(H5) virus clades 2.1.3.2a, 2.2.1, and 2.3.4 between 2013 and 2014. Viruses Other Than Influenza. 2015;9(5):271-276. Link

World Health Organization. Timeline of significant events related to H5N1 avian influenza. Link

Additional Reading Resources

Govorkova EA, Webster RG. Ongoing problems in influenza. 115–139 in Ann N Y Acad Sci., 2014; 1323. Link

Munster VJ, Wallensten A, Olsen B, et al. Influenza A virus patterns worldwide in untamed birds. 2006;312(5772):384–388 in Science. Link

Appendix C: Organizations and Contacts

Establishments:

The CDC, the Centers for Disease Control and Prevention,

https://www.cdc.gov

1800-CDC-INFO (1-800-232-4636) is the number to call.

Organization for World Health (WHO):

https://www.who.int

For contact details, refer to the regional offices.

The United Nations Food and Agriculture Organization (FAO):

https://www.fao.org

For contact details, refer to the regional offices.

Animal and Plant Health, United States Department of Agriculture (USDA)

ApHIS Inspection Service:

Website: www.usda.gov/aphis

Contact: For specific program contacts, visit the webpage.

The National Institute of Allergy and Infectious Diseases (NIH)

Diseases of Infection (NIAID):

Website: www.nih.gov/niaid

Contact: For specific program contacts, visit the webpage.

Hotlines:

The Emergency Operations Center (EOC) of the CDC:

Line number: 770-488-7100

WHO Emergency Helpline:

Hotline: For contact details, see the regional offices.

USDA-APHIS Crisis Helpline:

Please call 1-866-536-7593.

National Hotline for Animal Health Emergency **Assistance:**

Please call 1-866-536-7593.

Extra Sources:

Regional Health Authorities:

For information on avian influenza surveillance, prevention, and response initiatives in your area, get in touch with your local health department.

Veterinary Services:

See local physicians or veterinary clinics for advice on how to prevent and control avian influenza in livestock and pets.

Websites on Public Health:

For more information, go to the websites of the relevant research organizations, universities, and public health agencies.

www.ingramcontent.com/pod-product-compliance
Lightning Source LLC
Chambersburg PA
CBHW071925210526
45479CB00002B/553
* 9 7 9 8 3 2 8 8 9 8 3 9 3 *